CAREERS IN
CONSTRUCTION CONTRACTING

CHOOSING A CAREER SHOULD BE AN exciting experience. The first thing you need to do is ask yourself a few questions. What do you like to do? What are you good at? The idea is to find the career that answers *both* questions. Do you like to work with your hands? Can you keep track of a thousand details? Are you a good negotiator? If so, a career in construction contracting may be for you.

Construction contracting is a fundamental business. Almost everything requires some kind of construction. People live in houses and apartments, they work in office buildings and factories, they shop in small stores and enormous shopping malls, and they drive on roads to get there. Construction contractors build all of these things. Construction is so important to the overall state of the economy that the number of homes being built in a month – a statistic known as "housing starts" – can make or break Wall Street. Many huge industries make products primarily for construction, including lumber, concrete, gypsum, copper and steel.

Construction contracting is an all-encompassing business that includes trades ranging from carpentry to masonry, plumbing to electrical, steelwork to painting. Any of these trades are excellent avenues into the construction contracting business, and all are explored in detail in other Career Reports in this series. Although this report will touch upon these trades and a few others, it is mainly concerned with careers in construction management and what you will need to do to achieve that goal for yourself.

Construction is notoriously susceptible to fluctuations in the overall economy. When times are good, people and businesses expand by investing in new homes and commercial buildings. When times are not so good they tend to hunker down in the spaces they already have and wait for the economic storm to pass. The construction contracting business may grow rapidly for a few years, then flatten out or even dip a bit, and then grow again. Do not bother to try to time the job market as you plan your own future. There will always be opportunities.

WHAT YOU CAN DO NOW

YOU CAN GET SOME EXPERIENCE IN construction contracting right now. Find a part-time job with a contractor. Many contractors hire part-time workers to help out on jobs. You will not be doing highly skilled work like carpentry or plumbing, but you will be around contractors and tradespeople doing their jobs. You will learn from observing. You will probably be sweeping and vacuuming, hauling debris to the dumpster, and loading and unloading trucks. You will learn about quality control and construction-site safety, both of which are very important. Opportunities are best during the busy summer construction season.

Many construction contractors get their start in a trade, like carpentry or plumbing. Set your sights on a small construction project and figure out how to do it yourself. How about a playhouse for younger kids? A doghouse with lighting and windows? A backyard shed? There is nothing like hitting your thumb with a hammer to make you appreciate carpentry skills.

Construction managers often start out as skilled tradespeople, and need to have a good grasp of all the trades being used on their projects. But managers spend most of their time leading, planning and supervising. If you think you want to be a construction manager one day, you will need to find out if you have what it takes to be a leader. Take charge of something. Step up and become president of a club or team at school. Learn how to run a complex operation.

HISTORY OF CONSTRUCTION CONTRACTING

THE HISTORY OF CONSTRUCTION CONTRACTING parallels the history of architecture and engineering. By some accounts, modern humans have been around for as long as 250,000 years. For most of that time they lived in caves and temporary structures like lean-tos to protect themselves from the elements. Early humans did not know how to grow crops or raise livestock, so they moved constantly in search of food. They did not build permanent structures because they never stayed in one place long enough to need them.

That started to change in the Neolithic period, about 10,000 BC. The development of agriculture allowed human societies to settle in one place, creating demand for permanent structures. The lack of metal tools limited the types of building materials that could be used, but Neolithic humans did a good job with what they had. Dry-stone walling, for example, used human ingenuity to fit stones together so well that they did not need mortar. Some dry-stone walling structures still exist today, even after thousands of years of exposure to the elements. By the end of the Neolithic period, about 5,000 BC, humans had progressed to basic post-and-lintel construction and had begun to use timber as a major construction material. Although the timbers have long since disintegrated, archaeologists can pick out the holes where posts used to be by examining the soil around ancient construction sites.

Following the Neolithic era, humans started to use new construction techniques like bricks made from mud and straw baked in the sun. Bricks were not only sturdy and easy to make, they could also be manufactured in advance before they were needed. This is a critical development. The earliest buildings were all custom-made at a specific time and for a specific purpose. Bricks allowed societies to stockpile a basic building material that could be used at any time and for almost any purpose.

The ancient Romans made great strides in building technology, using advanced methods to construct buildings that can still be found scattered around Europe and North Africa today. The Romans knew enough about physics to build pulleys and cranes that allowed them to hoist building materials to great heights. They also made extensive use of lime mortar to bind building materials and to make concrete. The success of their methods is evident in buildings like the Pantheon in Rome, which still stands proudly and little changed. The Romans also perfected the idea of the state, the government entity responsible for public construction projects like roads, bridges, public buildings, and aqueducts (to carry water). Government contracting has been a major part of construction contracting ever since.

Construction techniques improved steadily during the Medieval and Renaissance eras, enabling greater sophistication and detail than ever before. Walls were supported by flying buttresses, and spans were wider. Finished lumber also became widely used during this period, as water-powered sawmills were able to turn logs into boards. Cities grew at a remarkable pace, creating new demand for public infrastructure.

Construction contracting did not always exist as we know it today. Architects and engineers were held in high esteem, as were skilled artisans like stonemasons and carpenters. The huge numbers of unskilled laborers required to build the pyramids, for example, were slaves whose only purpose was to provide muscle power. It could be argued that the standard labor practices we know today came into being during the Renaissance. Skilled laborers like carpenters, bricklayers and stone-masons were hired to construct buildings. Unskilled laborers were hired mostly on a daily basis, depending upon how much work needed to be done that day.

Projects were led by construction managers and supervisors, most of whom came up through a skilled trade. Projects tended to be directed by the same people from beginning to end, but there could be considerable turnover among the skilled and unskilled workers. Many of Europe's great cathedrals, for example, took more than 100 years to build. Some workers spent their entire working lives on a single project.

The next big advance was the use of iron as a

load-bearing material. First used in the early 18th century, iron could be worked into whatever shapes the builder needed and could bear significant stresses. The advent of iron in construction led directly to the invention of the skyscraper in the late 19th century. Featuring a steel-frame superstructure, the Home Insurance Company building in Chicago rose to an incredible 10 stories when it was constructed in 1885. Today, all tall buildings feature a steel superstructure.

Residential construction evolved over the centuries, too. Only a few hundred years ago, most people lived in tiny huts or small spaces inside large apartment buildings. Rural-dwellers built their own homes, often huts, shacks or cabins, while city-dwellers packed together in large tenement buildings consisting of many cramped units.

In the United States, the vast expansion of suburbia after World War II gave construction contracting a vast new market for single-family houses. Most single-family houses today are built by large developers who buy parcels of land and build dozens or hundreds of houses at the same time. These projects are called subdivisions. Developers may hire hundreds of individual contractors to build a subdivision, and the projects can take years. There is also demand for contractors to build custom houses one at a time for individuals, and to renovate existing houses.

The commercial side of construction contracting features projects large and small. Many contracting companies specialize in building and maintaining standard commercial buildings like strip malls, gas stations and restaurants. Others deal mostly in very large projects like enclosed shopping malls or very large office or mixed-use buildings. From a career perspective, building single-family houses or small commercial buildings requires essentially the same set of skills. From the business perspective, building houses and small commercial buildings requires much less capital investment than building major commercial buildings. Small contracting companies have been established with nothing more than an old pick-up truck and a few hundred dollars in tools. The companies that build skyscrapers are multimillion-dollar enterprises.

Since ancient times governments have been important sources of construction contracts. Roads, bridges, aqueducts, sewer systems and public buildings are mostly built by construction contractors. This is known as "heavy civil" construction. Governments tend to maintain their own infrastructure, but they rarely build it themselves. Governments do not keep a big construction staff employed all the time. When a building or road is done, the government takes over the maintenance, and the construction contractors move along to other projects.

WHERE YOU WILL WORK

WHEREVER YOU GO TO ESTABLISH your career, you should be prepared to stay for a while. Construction contracting is an intensely local business. Contractors make names for themselves within their markets and become very familiar with local building codes and architectural styles. Even in the age of the Internet, word-of-mouth remains the favored way for contractors to get new business.

For example, the fast-growing suburban fringes of major metropolitan areas offer many opportunities to get into construction, and especially residential and retail construction. Suburban areas tend to be carefully planned, with city and county governments plotting every square inch for what is known as the "highest and best use." That means they set aside land for residential development, retail development and office and industrial development. That way people can live, shop and work within a relatively small area. Suburban areas tend to contain mostly houses, apartments, and retail buildings for shopping. They may also have some office space and light industrial development, generally concentrated in what are called "industrial parks."

Contracting in densely populated cities requires different expertise. Where suburbs tend to build out, cities tend to build up. That is because land is generally more expensive in cities than in wide-open suburbs, and because it is very hard to find large parcels of land available for sale and development in cities. In the suburbs, an acre of land may host two or three houses. In a city, an acre may support a tall building containing hundreds of apartments or condominiums.

Rural construction comes with its own challenges and opportunities. Single-family houses are the norm in small towns and the countryside. Agricultural buildings like barns and silos may not be as complicated as skyscrapers, but they also require specialized knowledge. Small towns may have one or two contracting companies who get all the contracts. Breaking into a small established market can be very difficult.

DESCRIPTION OF THE WORK

Residential Construction

Residential construction is the most common specialty. Residential contracting companies may be very small, consisting of fewer than 10 full-time employees. Even the largest, however, may not have more than 100 or so full-time employees because they hire subcontractors to do most of the work.

Constructing a building requires the skills of many people for a relatively short period of time. When the project is finished most of those people are no longer needed. That is why contracting companies use subcontractors. A general contractor may hire a plumbing subcontractor to handle the plumbing for a particular project. That plumbing work may require five days of labor, split into three at the beginning of the project and two near the end. In between, the plumber is free to work for other general contractors on other projects.

Many construction supervisors work their way up in the residential contracting business. They start out with summer jobs in high school, and graduate to apprenticeships. They need to learn about many aspects of the work, like project management, cost estimation, regulatory compliance, labor relations, and environmental regulations. Many earn bachelor's degrees in construction science, architecture or engineering and then get on-the-job training after they graduate. Residential construction companies may build one house at a time for individual clients, or develop entire subdivisions, consisting of hundreds of dwellings. They may concentrate on one project at a time, or have several going so they can keep their subcontractors busy.

One job that all residential general contractors do is deal with homeowners. This is a fundamental difference between residential and commercial contracting. Commercial contractors generally deal with professionals who know what they want and know how the business works. Homeowners often do not have a clue about construction, and have no idea what it entails or how long it realistically takes. Sometimes they do not really even know what they want, and argue with contractors when they try to present reasonable options. Mostly homeowners are pleasant and understand that contractors are not magicians, but every contractor has a few stories to tell about nightmare homeowners.

Residential contracting offers opportunities for entrepreneurship. It does not take much capital to start a small business specializing in something like laying tile or installing cabinets. Over time, and with experience and contacts, that can grow into a general contracting business, building entire residences.

Renovation

Many construction contractors specialize in renovation projects. Houses and apartment buildings need repairs, expansion, and new finishes, from time to time. They may need a complete overhaul, from top to bottom, even a "gut rehab," where everything is torn out except the exterior walls. Sometimes residents just decide they need a little extra space and engage a contractor to make the basement livable or add a room out back.

Kitchens and bathrooms are especially lucrative for specialty contractors. Nothing increases the value of a home quite as much as renovating a kitchen. Homeowners will pay $25,000 or more to renovate their kitchens or $10,000 for a new bathroom. Garages are another popular specialty. Many people want their garages to be more than just places where they park their cars. They want to have room for cars, bicycles, motorcycles, lawn equipment, tools, and workbenches.

Commercial Construction

Many contractors have the resources to build small commercial buildings like strip malls and small office buildings. Even contractors who specialize in residential construction sometimes take on small commercial projects. Relatively few contractors, however, have the means to build skyscrapers.

Commercial contractors tend to deal with business owners and investors who know what they want and have a good understanding of construction and finance.

They have lawyers and accountants go over their contracts before they sign anything, and they expect contractors to stick to schedules.

Industrial Contracting

Industrial construction is concerned with building factories, warehouses, refineries, plants, mills, pipelines, and other structures used for industrial production. Some industrial construction is fairly simple, consisting of open-plan buildings that can be used for machinery or small assembly operations. Other industrial construction is extremely complex.

Government Contracting

Government construction contracting includes residential, commercial, industrial and heavy civil projects. Governments at all levels constantly build and update their facilities – city halls, streets, military base housing, runways, office buildings, bunkers, bridges, parking lots, sewers, libraries, schools.

Building for the government is essentially the same as building for any other client. The contracting process itself is typically much more complicated. Governments identify a need and advertise a contract. Qualified contracting companies bid to do the work and the government usually has to go with the lowest bidder. This is the case with simple projects like laying sidewalks or repairing existing infrastructure. For complex jobs, governments typically publish what is called a "request for proposal," or RFP. An RFP identifies a need but does not come with a specific plan. Contractors not only have to bid for the contract, they have to present their own plans. They win or lose based on their price and how well the plan fits the need.

Government contracting is also subject to many steps not generally required in the private sector. Major projects are subject to public hearings, so taxpayers can weigh in on them. They are also subject to several layers of review by various committees and experts. Politics and funding can start or stop a project in an instant. There are also diversity requirements for most government contracts.

STORIES OF WORKING CONTRACTORS

I Am a Residential Construction Contractor

"The company I work for has been around a long time. Over the last century or so we have built thousands of houses throughout our metropolitan area, in the city and its suburbs. We do some commercial construction, and build a small apartment building from time to time, but our bread and butter has always been houses.

I got into this career while I was still in high school. I needed a summer job and a friend of a friend lined me up with a construction contractor. Mostly I showed up early to unload trucks, and then spent the day doing miscellaneous chores. I was definitely an unskilled laborer, but I learned all about the skilled work just by being around it. The boss liked me. I showed up on time, didn't mind working late, and was always willing to learn something new.

When I came back for a second summer I got to work alongside the carpenters. I liked working with wood, and got a real kick out of seeing a pile of boards turn into a house. This informal apprenticeship turned into a formal, four-year apprenticeship after I graduated from high school. My program required 2,000 hours of work experience per year and 144 hours of technical training. It was a rigorous program and I learned a ton.

Some of the carpenters I worked with had been in the business for longer than I had been alive. They knew everything. Even after four years I knew that I still had a lot to learn.

I worked as a journeyman carpenter for several years, perfecting my trade and learning more about the construction business. I always stepped up to lead projects and teams and eventually landed the position of construction supervisor.

Construction supervisors are the on-site leaders for every project. They take their orders from the company's owners or senior managers, and then carry them out at construction sites. For most workers, the construction supervisor is the only boss they'll ever see.

This is a very complicated job. It's not enough to know everything about carpentry, which is my specialty. I have to know something about plumbing, electrical, windows, heating and air conditioning, painting, roofing, concrete, tile, hardwood flooring, and landscaping. I have to make sure that I schedule everybody in the right sequence, and that everybody has what they need when they arrive on the job site and enough time to complete each task.

Construction projects rarely work out exactly according to the plan. To give an example, I recently completed a project that was delayed by a month because of a leaky pipe. A plumber had some difficulties installing plumbing in a bathroom in a second-story renovation. None of the other trades could get into the site until the plumbing was buttoned up. A two-day delay in the plumbing had a huge ripple effect because the electricians had to put off their work, but two days later the electricians were already scheduled to be at another job. The drywall installers couldn't do their part until the electricians were done, but the electricians couldn't get back to the project for a week. By that time, the drywallers were already booked that week. Then a month went by. Most construction companies have very small full-time staffs but have connections with all the different subcontractors, who work for many different general contractors. You can't always get them right when you need them.

One of my most important functions is making sure that all the work we do is in compliance with local building codes. There are codes for everything, from the width of doors and hallways, to the kind of materials that have to be used for specific applications.

I earned a bachelor's degree in business administration part time, as I moved up the ladder with the company. I still do some carpentry, and like to think of myself as a "working" construction supervisor. When I have the time I strap on a tool belt like the rest of the crew and get to work.

As a partner in the company I also spend time serving the community in various ways. I am on the board of the local chamber of commerce, for example, and do charity work in the community."

I Am a Construction Contractor Specializing in Kitchens and Bathrooms

"Do you know the easiest way to add value to a home? Update the kitchen and bathrooms. Kitchen and bath renovation is a specialty that is always in demand, and there is very good money to be made.

I started out in this business as a cabinet installer, part time while I was in college. I majored in business and had always assumed that I would go into a career in finance. I grew to like the cabinet business and never left.

When I started out, I worked full time alongside an established cabinet installer for a summer, and then part time for about a year while I finished college. After I figured I had learned enough to do the job on my own I started freelancing. I bought a van and loaded it up with all the tools I would need, from small tools like hammers and pliers, to major power tools like a table saw and a compound miter saw. After graduating from college, I kept at it and my business grew rapidly. Within a year I had to hire an assistant. A few years after that I had to buy additional vans and hire teams of installers to be in several different places.

Running any business is all about achieving maximum efficiency in order to make the most profit. Right now I have four full-time teams installing kitchens and baths five days a week – a busy schedule. I also hire subcontractors to help out when things get very busy.

I get a kick out of the way that good kitchen and bath designs come together. It's really quite remarkable that a dozen or more cabinets, some granite, a few appliances and assorted bits of lighting and plumbing can fit together so well and result in something so useful and aesthetically pleasing."

I Am a Residential Property Developer

"I'm not a contractor but I hire many of them. I am a residential real estate developer. I build subdivisions consisting of anywhere from a few dozen houses to a few hundred. I may have dozens of construction contracting companies working on my projects at one time.

A subdivision is a parcel of land that has been divided into smaller parcels to be put to specific uses. If I buy 100 acres of farmland on the edge of town I have to ask the local government to re-zone the land for residential use. Usually they accept my recommendation, but not always. Next, I have to submit a land-use plan to the local jurisdiction, usually a city or county government. The land-use plan will show where all the houses, yards, streets and parks will go, along with necessities like drainage ditches and wetlands mitigation areas. If I start out with 100 acres I can generally end up with 50 homes on it.

After all the plans have been approved I start the process of hiring contractors. The construction contractors I employ are usually general contractors. That means they are in charge of the project, and they hire subcontractors to do each trade job. I may hire multiple general contracting companies and assign each company to one block of the subdivision, for example.

A new subdivision development can support several hundred jobs for a year or more, depending upon the size. I have the same scheduling headaches regular construction contracting companies do, but on a larger scale. People are often amazed at how quickly subdivisions can rise from former farmland.

I got into this business as a small-time real estate investor. I started flipping houses in my 30s to make a little extra money. I would buy a house, remodel it, and then sell it at a profit. I did good work and built a solid reputation with local contractors and with the banks that financed my projects. When the time came to take a big risk on an entire subdivision, the bankers and the contractors were behind me.

If I were to give one piece of advice to anybody pursuing any real estate business, it would be stay honest, do your best to stick to your schedules, and always be willing to do a favor or two. This is an intensely local and personal business. Your most precious asset is your reputation."

I Build Road and Highway Infrastructure for Governments

"Most governments don't build their own infrastructure, they use a company like mine to build it. I specialize in heavy civil construction, especially road and highway infrastructure.

I am a senior construction supervisor for a very large company. We build much of the infrastructure in a major metropolitan area. We maintain constant contact with city, county, state and federal government offices. We can build simple residential streets, urban corridors with four lanes and complex sewage and drainage systems, bridges and tunnels. We have a fleet of trucks and hundreds of full-time employees. During the busy summer construction season, we can have more than 1,000 subcontractors working on our projects.

What we do is very different from building single-family houses. We pour concrete by the thousands of tons and buy almost enough steel in a typical year to build an aircraft carrier. We employ engineers, architects and geologists.

A major highway project can take years or even decades to complete. The logistics are very complex. Trucks bring raw materials to the work site, but they usually have to stick to a predetermined route because they are so heavy they can't use some streets. The work site moves a few feet each day. Politicians alter contracts and appropriations even after they have been signed, resulting in costly delays. Environmental regulations are constantly varying. Community groups file lawsuits because they don't like the noise or the idea of additional traffic in their areas. All part of my job.

I like my work because I appreciate the importance of what we do. The first highway built across the United States opened in 1913 – not really so long ago. The infrastructure we take for granted enables us to go about our daily lives with a degree of speed and efficiency unknown to earlier generations. I am proud of my contribution."

PERSONAL QUALIFICATIONS

BY THE TIME YOU BECOME A CONSTRUCTION manager or supervisor, you will know about most aspects of building. Even if you rise up through the ranks by pursuing one trade specialty, you will be knowledgeable about all the other trades. You will need to be able to schedule all the different trades, and to make sure that they have enough time and supplies when they are on the job.

Construction contracting is not a career where you learn everything you need in school. Building codes and other legal requirements change constantly. Architectural styles and tastes vary year by year. Technology and equipment improve. There will be classes and seminars, but mostly there will be endless opportunities to learn on the job. On-the-job learning is the best, most challenging education you will ever get.

You have probably heard the term "multitasking," describing the process of accomplishing multiple tasks at the same time. It is a necessary skill if you want to succeed as a construction contractor. On a typical day at a construction site, a construction supervisor could be directing the work of plumbers, electricians, painters, and carpenters all at the same time. The supervisor will also be keeping track of the number of feet of pipe used by the plumbers, the type of receptacle boxes used by the electricians, the gallons of paint used by the painters, and the board-feet of lumber used by the carpenters. The supervisor will also have to track the number of hours of labor put in by all the workers on the site, and be sure that pay and working conditions follow rules set by unions, local governments and the federal government. The supervisor will then have to deal with the clients and employers, and make sure that they are satisfied at each stage in the process. If the supervisor misses any of these steps, there could be chaos.

Construction contractors are good businesspeople. They negotiate complex contracts with clients and subcontractors. As projects progress, the supervisor has to make many small decisions about buying materials and scheduling labor that will have an impact on profit. Contractors do their best to make accurate predictions, but things have a tendency to drift when a project goes on for several months. It is the boss's job to keep the project on course.

ATTRACTIVE FEATURES

CONSTRUCTING OR RENOVATING buildings can be very expensive. Many skills and trades are involved, raw materials are costly, and there are innumerable rules to be followed. That is why people hire professional contractors even for relatively simple jobs. Construction contractors are well paid for what they do, and they earn it. The work is complex, sometimes dangerous, subject to significant liability, and difficult and expensive to change once it is completed.

Construction managers who work as employees for large

companies can earn more than $100,000 per year, depending upon their specialty. Self-employed contractors who own their own business can earn millions of dollars per year.

Opportunities for entrepreneurship are exceptionally good. In fact, many construction contractors are self-employed for their entire careers. They start out as skilled tradespeople working as subcontractors for contracting companies. They set themselves up as sole proprietors to run their business, even though their business might consist of only one person. As time goes by, they expand their subcontracting operation by taking on a few assistants. Eventually they find themselves in charge of a real business with employees, inventory, office space, and a fleet of vehicles. Multiple subcontractors sometimes merge their operations into one business, creating a full-service contracting company. Success in the construction contracting business requires dedication, a willingness to take calculated risks, and meticulous attention to detail. You may be attracted to this business because you are developing a skill in a building trade. To become the owner of a full-service contracting company, you have to want to be the boss.

There is a great deal of satisfaction to be had in pursuing a career in something as basic as construction contracting. You will build and renovate the homes and workspaces that people use every day. You will be an essential part of your community.

Contractors are often heavily involved in local institutions like chambers of commerce and other civic organizations. Their livelihood depends upon keeping their name front-and-center in the community they serve. They sponsor Little League baseball teams, for example, and do charity work for organizations like Habitat for Humanity and the Veteran's Administration. If you like the idea of being a high-profile leader in your local community, this career may be a good fit for you.

UNATTRACTIVE ASPECTS

THERE ARE SIGNIFICANT DOWNSIDES you should consider. This is hard work, fraught with rules, regulations and liability. Few industries are as highly regulated as construction contracting. Almost all trades have to be licensed and certified. That means that if you hire plumbers you need to make sure that their licenses are up-to-date and that they have the proper certifications to do the work you need them to do. Failing to do so could leave you open to serious liability if something goes wrong. Liability is a major issue with contracting. General contractors are typically liable for major structural issues but pass liability for specific issues like plumbing and electrical on to their subcontractors. New construction and renovations usually come with warranties of anywhere from one to three years. If a contractor puts an addition on a house and gives it a one-year warranty and it burns down six months later due to an electrical fault, the electrician should be liable. Such mistakes cost money.

Liability and regulation are necessities because construction contracting can be a dangerous business. Even the humblest home renovation involves numerous power tools, and any contractor can tell a million stories about close calls with them. Minor injuries are common, from cuts and bruises that are mostly annoying to issues like back injuries that can sideline a worker for weeks or even months. State and local regulations dictate safety requirements at construction sites, along with the federal Occupational Safety and Health Administration, or OSHA. Complying with the numerous rules and regulations is expensive and time-consuming.

Construction contracting tends to be a very seasonal business. Building is much easier in warm weather than in cold weather. Extreme cold can make construction impossible. Many unions require workers to be paid more if they have to work in extreme conditions, whether cold or hot. Clients generally will not start a renovation project in winter if it involves opening up a building to the cold. Obviously, this phenomenon is less pronounced in the South, where winters are fairly mild, but the overall trend is for construction to peak in the summer and fall off for a few months during the winter. Weather of all kinds can affect construction. A day of rain can force subcontractors to reschedule multiple jobs and create a ripple effect of delays. High winds can destroy work in progress.

EDUCATION AND TRAINING

THERE ARE MANY WAYS TO GET THE education and training you will need. Many contractors simply work their way up through the ranks until they can take the lead. Today, however, most earn bachelor's degrees in a relevant discipline. Both paths have their advantages and disadvantages.

Traditionally, most construction contractors started out as unskilled laborers, learned a skilled trade, established a reputation, and eventually broke into the managerial ranks. This was especially true for general contractors specializing in residential and basic commercial construction. To take this route, the first thing you will need to do is learn a skilled trade like carpentry, plumbing or electrical. Some trades require that you go through an apprenticeship in order to become certified. In an apprenticeship you work alongside an established pro for a specified period of time, usually a year or two. Some classes and examinations may also be required. Requirements may be dictated by union work rules or by state law. Satisfying a union requirement may or may not be sufficient for a state or local licensing requirement.

Arrangements vary from one jurisdiction to the next.

Keep in mind that the careerists who move up through the ranks without any formal education beyond high school are not likely to be chosen to build skyscrapers, nor are they likely to build a large business able to serve more than one community. These endeavors require serious engineering expertise, and running any sizable business requires advanced business knowledge. If you really want to make it to this level, you will need to go to college.

Many self-taught construction contractors earn bachelor's degrees in business administration later in life when they realize that they need to enhance their skills in order to grow their business. There are other options. Recognizing that many contractors start out in the trades and have a great deal of real-world experience, many community colleges offer two-year associate degrees in construction technology and construction management. These programs stress basic business skills like management and accounting, as well as construction skills like cost estimating and regulatory compliance. Many contractors also earn certification from the Construction Management Association of America or the American Institute of Constructors.

Many colleges and universities also offer bachelor's degree programs in construction science and construction management, and degrees in architecture, engineering or business administration will always be helpful to a career in construction contracting. Earning a degree will supply you with the advanced knowledge necessary to grow your business, both by expanding your existing services and by taking on new ones. A degree may also be necessary to get a job with a large contracting firm, like the ones that compete for huge commercial buildings and public works projects.

Will you need a master's degree? You can bet that the architects and engineers who design and build skyscrapers have advanced degrees, and so do the construction managers and supervisors who lead such ambitious projects. Many managers at small contracting companies go back to school later in life to earn master's degrees in business administration. You may find that you will need to return to the classroom someday, although this is not a decision you have to make right now.

EARNINGS

CONSTRUCTION CONTRACTING CAN produce very high incomes. Skilled tradespeople are often paid according to a standard scale enforced by unions at the local level. Scales vary according to region, with tradespeople in New York City earning more than their counterparts in Memphis, Tennessee. Electricians, for example, can earn anywhere from $20 to $60 per hour depending upon their region and other factors like seniority and the need for any special licenses or certifications. Plumbers can earn as much as $100 per hour. General carpenters can earn $20 to $40 per hour, while trim carpenters earn a bit more for detail work. Pay for tradespeople usually includes mandatory benefits like health insurance and programs like Worker's Compensation. These fringe benefits can be worth almost as much as the hourly pay.

The median salary for construction managers is about $85,000 per year. Careerists just out of college can expect to start at about $45,000 a year. Construction managers leading large projects for major firms can earn more than $150,000 per year. The biggest variable is project complexity. Construction managers who lead the construction of downtown skyscrapers are paid more than their peers turning out single family houses in the suburbs. They have the advanced degrees and years of experience necessary to compete for the top jobs.

As with most professions, the highest earnings go to the entrepreneurs who own their own businesses. Many construction contractors are self-employed for their entire careers, first as freelance tradespeople and later as independent business owners. It is difficult to estimate average earnings for construction contractors who own their own companies, but they could reach $1 million or more in a very productive year.

OPPORTUNITIES

GROWTH IN THE CONSTRUCTION industry is expected to keep pace with the overall economy for the foreseeable future. Contractors with degrees are in high demand for jobs with major contracting companies. Many contracting companies prefer to hire well-educated careerists for their managerial positions. Construction management demands a full array of skills, like business administration, accounting, and a solid understanding of architecture and engineering. Even if you work your way up through the ranks in a small company that deals mostly with residential construction, you will still need to earn a degree if you want to break into the world of large-scale commercial construction and public works.

Targeting a specialty can also be helpful. The investors who pay for construction projects think in terms of cost per day. Speed is essential to keeping costs down. A client who wants a restaurant, for example, will pay for a specialist in building restaurants. There are entire firms dedicated to a single building specialty, like hospitals or parking garages. Specialists already know all the details, and they are likely to finish projects faster and more reliably. This can be said for most professions, but construction contractors were traditionally known for being able to solve problems on the fly and "making it up as you go along." Being able to think on your feet is still a valuable skill for a construction manager, but specialists are in greater demand than ever before.

One of the specialties most in-demand is construction contractors who know how to retrofit buildings to make them more energy efficient. Contractors who earn LEED certification (Leadership in Energy and Environmental Design) from the US Green Building Council can compete for lucrative retrofit projects to bring old buildings into the modern era. LEED certification has become extremely popular in recent years, as one of its aims is to reconcile the public desire to preserve old buildings with the desire to use less energy. Retrofitting old buildings with new heating and air conditioning systems, insulation, windows and other features can give them a new lease on life.

GETTING STARTED

WHEN THE TIME COMES TO GET started on your first full-time job you should be well prepared. Get your personal marketing materials in order. Even if you find your way into a position through personal connections, you should always be ready with an up-to-date résumé. Your current employer may need one on file. If you are not confident in your ability to write a résumé do not hesitate to seek out some assistance. There are innumerable books and software applications that can help you to put together a proper résumé, and your school may offer résumé-writing classes or services to help you out. You should have a properly formatted printable résumé even if all of the job applications you fill out are online. Having a résumé ready to go makes it possible to cut and paste information. This also ensures consistency from one application to the next.

Many jobs are never advertised. Often, employers simply want to get a particular person on the payroll and assign specific duties later. If you have completed an apprenticeship and have some work experience, there may be an employer who will hire you given the chance. Many new careerists find their first full-time jobs with companies where they have worked part time or over the summer, or where they completed an internship. To the employer, you are someone who already understands the company culture and the nature of the work. By hiring you they save money in advertising, interviewing and training a completely new hire. Even if your old connections do not have a position for you they may know somebody who does. You might be surprised at how helpful former employers can be if you left them on good terms.

No matter how you go about your job search remind yourself that this is your first job. You are not committed to it for life. It does not have to be your dream job. It just has to get you into the construction contracting industry and give you some skills, experience and credibility. The details are not very important when you are starting out. There are many years ahead to change course if necessary. You might even discover something new along the way. Be sure to update your résumé from time to time. Putting your new skills and experience down on paper can be a very uplifting experience.

ASSOCIATIONS
PERIODICALS
WEBSITES

■ Access Board
www.access-board.gov

■ Advisory Council on Historic Preservation
www.achp.gov

■ **American Council for Construction Education**
www.acce-hq.org

■ **Angie's List**
www.angieslist.com

■ **Army Corps of Engineers**
www.usace.army.mil

■ **Associated Builders and Contractors**
www.abc.org

■ **Associated General Contractors of America**
www.agc.org

■ **Association of Union Constructors**
www.tauc.org

■ **Building Design and Construction**
www.bdcnetwork.com

■ **Building Trades Association**
www.buildingtrades.com

■ **Columbia University School of Continuing Education**
www.ce.columbia.edu

■ **Construction Industry Compliance Assistance Center**
www.cicacenter.org

■ **Department of Commerce**
www.commerce.gov

■ **Department of Housing and Urban Development**
www.hud.gov

■ **Department of Transportation**
www.dot.gov

■ **Engineering and Construction Contracting Association**
www.ecc-conference.org

■ **Environmental Protection Agency**
www.epa.gov

■ **Equipment World**
www.equipmentworld.com

■ **General Building Contractors Association**
www.gbca.com

■ **General Services Administration**
www.gsa.gov

■ **International Brotherhood of Electrical Workers**
www.ibew.org

■ **International Union of Operating Engineers**
www.iuoe.org

■ **National Association of Home Builders**
www.nahb.org

■ **National Association of Women in Construction**
www.nawic.org

■ **National Center for Construction Education and Research**
www.nccer.org

■ **National Electrical Contractors Association**
www.necanet.org

■ **National Utility Contractors Association**
www.nuca.com

■ Paxton Patterson Education
www.paxtonpatterson.com

■ United Association
www.ua.org

www.ingramcontent.com/pod-product-compliance
Lightning Source LLC
Chambersburg PA
CBHW070755180526
45168CB00004B/1628